PB99-917005
NTSB/HZM-99/01

NATIONAL TRANSPORTATION SAFETY BOARD

WASHINGTON, D.C. 20594

HAZARDOUS MATERIALS ACCIDENT REPORT

FIRE AND EXPLOSION OF
HIGHWAY CARGO TANKS, STOCK ISLAND
KEY WEST, FLORIDA
JUNE 29, 1998

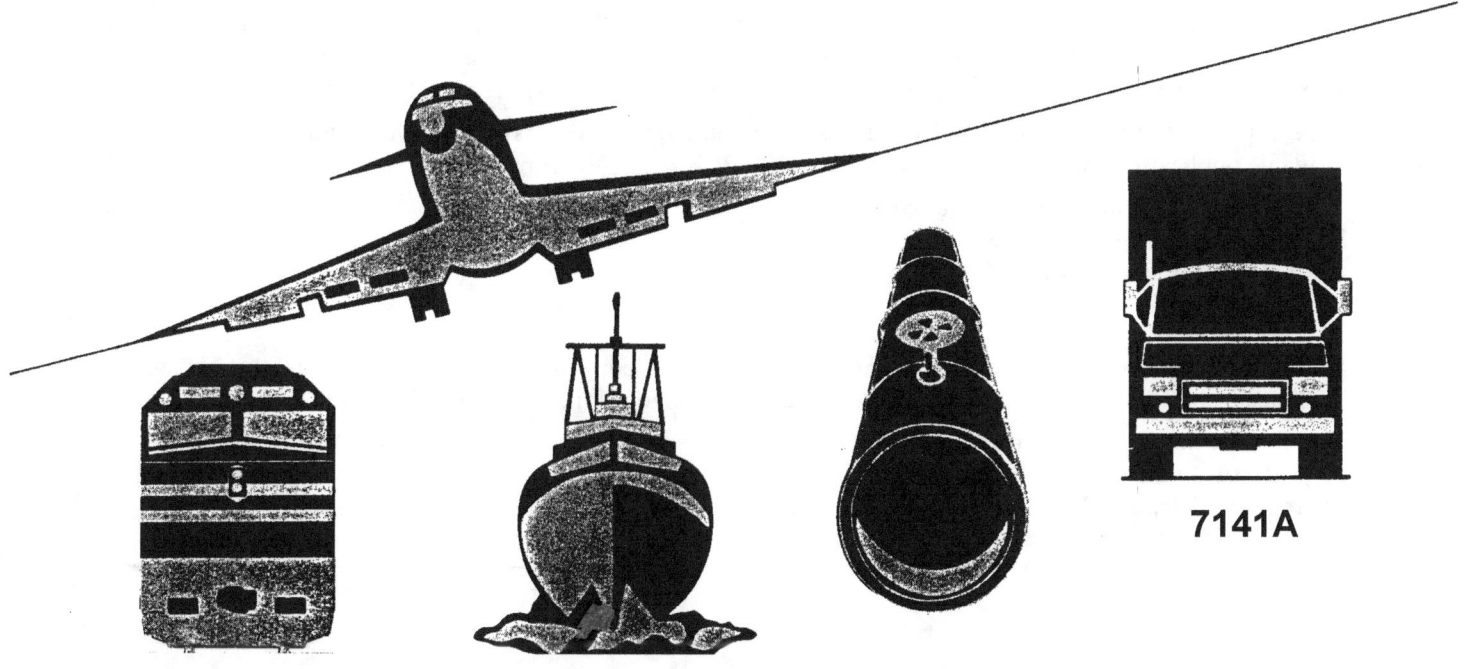

7141A

National Transportation Safety Board. 1999. *Fire and Explosion of Highway Cargo Tanks, Stock Island, Key West, Florida, June 29, 1998.* Hazardous Materials Accident Report NTSB/HZM-99/01. Washington, DC.

Abstract: About 5:14 a.m., eastern daylight time, on June 29, 1998, at Stock Island, Key West, Florida, a Dion Oil Company (Dion) driver was on top of a straight-truck cargo tank checking the contents of its compartments and preparing to transfer cargo from a semitrailer cargo tank when explosive vapors ignited within the straight-truck cargo tank. The ignition caused an explosion that threw the driver from the top of the truck. The fire and a series of at least three explosions injured the driver and destroyed the straight truck, a tractor, the front of the semitrailer, and a second nearby straight-truck cargo tank. Damage was estimated at more than $185,000.

The safety issues discussed in this report are the adequacy of Dion's product-transfer procedures and training, the adequacy of the Federal Highway Administration's oversight of motor carriers' procedures and training for loading and unloading hazardous materials, and the adequacy of Florida's oversight of the fire safety of storage tanks.

As a result of its investigation, the National Transportation Safety Board issued recommendations to the Federal Highway Administration, Dion, the Florida State Fire Marshal, the Florida Department of Transportation, the Florida Department of Agriculture, the Florida Department of Environmental Protection, the National Fire Prevention Association, the National Association of State Fire Marshals, and the International Association of Fire Chiefs.

Hazardous Materials Accident Report

**Fire and Explosion of
Highway Cargo Tanks, Stock Island
Key West, Florida
June 29, 1998**

NTSB/HZM-99/01
PB99-917005
Notation 7141A
Adopted September 10, 1999

National Transportation Safety Board
490 L'Enfant Plaza, S.W.
Washington, D.C. 20594

this page intentionally left blank

Contents

this page intentionally left blank

Executive Summary

About 5:14 a.m., eastern daylight time, on June 29, 1998, at Stock Island, Key West, Florida, a Dion Oil Company (Dion) driver was on top of a straight-truck cargo tank checking the contents of its compartments and preparing to transfer cargo from a semitrailer cargo tank when explosive vapors ignited within the straight-truck cargo tank. The ignition caused an explosion that threw the driver from the top of the truck. The fire and a series of at least three explosions injured the driver and destroyed the straight truck, a tractor, the front of the semitrailer, and a second nearby straight-truck cargo tank. Damage was estimated at more than $185,000. As a result of its investigation of the accident, the National Transportation Safety Board identified three major safety issues:

The adequacy of Dion's product-transfer procedures and training.

The adequacy of the Federal Highway Administration's oversight of motor carriers' procedures and training for loading and unloading hazardous materials.

The adequacy of Florida's oversight of the fire safety of storage tanks.

The Safety Board determines that the probable cause of the accident was Dion's lack of adequate procedures and driver training, resulting in the driver's pouring a mixture of gasoline and diesel fuel from a plastic bucket into a cargo-tank compartment that contained a mixture of explosive vapors.

As a result of its investigation of this accident, the Safety Board makes recommendations to the Federal Highway Administration, Dion, the Florida State Fire Marshal, the Florida Department of Transportation, the Florida Department of Agriculture, the Florida Department of Environmental Protection, the National Fire Prevention Association, the National Association of State Fire Marshals, and the International Association of Fire Chiefs

this page intentionally left blank

Factual Information

Accident Narrative

In January 1995, Dion Oil Company (Dion) began renting property at Robbie's Marina in Stock Island, Florida, as parking space for Dion's fuel trucks. Dion subsequently began keeping a semitrailer cargo tank on the site as a temporary storage tank for fuels (gasoline and diesel fuel). The fuel in the temporary storage tank was transferred to straight-truck cargo tanks as needed for delivery to clients.

The driver injured during the accident had been on vacation from June 17 through June 27, 1998. Records indicate that he returned to work on June 28, the day before the accident, and delivered 400 and 280 gallons of premium unleaded gasoline to Sherman's Waterfront and to the U.S. Coast Guard Group Key West, respectively.

According to the driver, he arrived at Robbie's Marina shortly before 5 a.m., eastern daylight time, on June 29. He intended to transfer a load of diesel fuel from the temporary storage tank to his cargo tank. He stated that he had just arrived and was carrying a plastic bucket of mixed fuels that he had retrieved from under the temporary storage tank. He indicated that he believed the bucket contained a mixture of gasoline and diesel fuel that had spilled from hoses or fittings during previous cargo transfers.[1]

He climbed to the top of his vehicle, carrying the bucket, and opened the three compartment lids on his vehicle to determine the type of fuel each compartment held. He indicated that because the two back compartments opened without releasing pressure, he believed they held diesel fuel and that because the front compartment released pressure when it opened, he believed it held gasoline. He stated that he may have been pouring the contents of the bucket into the front compartment when he saw flames coming from the compartment and was thrown from the top of the truck.

Emergency Response

About 5:14 a.m., two deputies from the Monroe County sheriff's department were driving past Robbie's Marina when one of the deputies heard a loud hissing sound coming from the area of the cargo tanks, followed shortly afterward by several explosions and a fire. (See figure 1.) The deputies immediately radioed their dispatcher and reported the accident. At the same time, the owner of the marina was awakened by an explosion and went to the scene. The owner saw the driver lying about 75 feet away from his truck and pulled him away from the fire toward the marina. The driver had second- and third-degree burns on his hands, arms, and legs and a broken left knee.

---

[1]The driver stated that he disposed of the spilled material in the buckets under the temporary storage tank as part of his daily routine.

Figure 1. Fire involving the cargo tanks (Photo by Scott Wallace.)

About 5:15 a.m., the Monroe County dispatcher notified the Stock Island Volunteer Fire Department Chief, the Key West International Airport Fire Department, the Big Coppitt Key Volunteer Fire Department, and the Monroe County Ambulance Service. The Stock Island fire chief had been at his residence, which was near the marina, and arrived within minutes. He became the incident commander, and by 5:20 a.m., he had established an incident command post on Shrimp Road, the road that ran past the marina.

At 5:24 a.m., a Monroe County ambulance arrived but could not get to the injured driver because the road ran past the burning cargo tanks. The marina owner transported the driver in a private vehicle to the waiting ambulance. The ambulance took the driver to the Lower Florida Keys Health System hospital. (He was later transported by helicopter to Jackson Memorial Hospital Burn Center in Miami, Florida. He was in the hospital and unable to talk for almost 3 months. Once he was able to talk, he could not remember making deliveries the day before the accident and was not clear about all of the events on the morning of the accident.)

At 5:30 a.m., the incident commander asked the Boca Chica Naval Air Station Fire Department for help, and it sent a crash truck, which arrived at 5:42 a.m. As soon as it arrived, it began suppressing the fire with foam. By 6:15 a.m., the fire was sufficiently extinguished to allow the fire departments to begin search operations. At 11:09 a.m., all emergency response operations were discontinued.

Vehicle Information

Three vehicles, all owned by Dion, were involved in the accident. Figure 2 shows where the vehicles were when the accident happened.

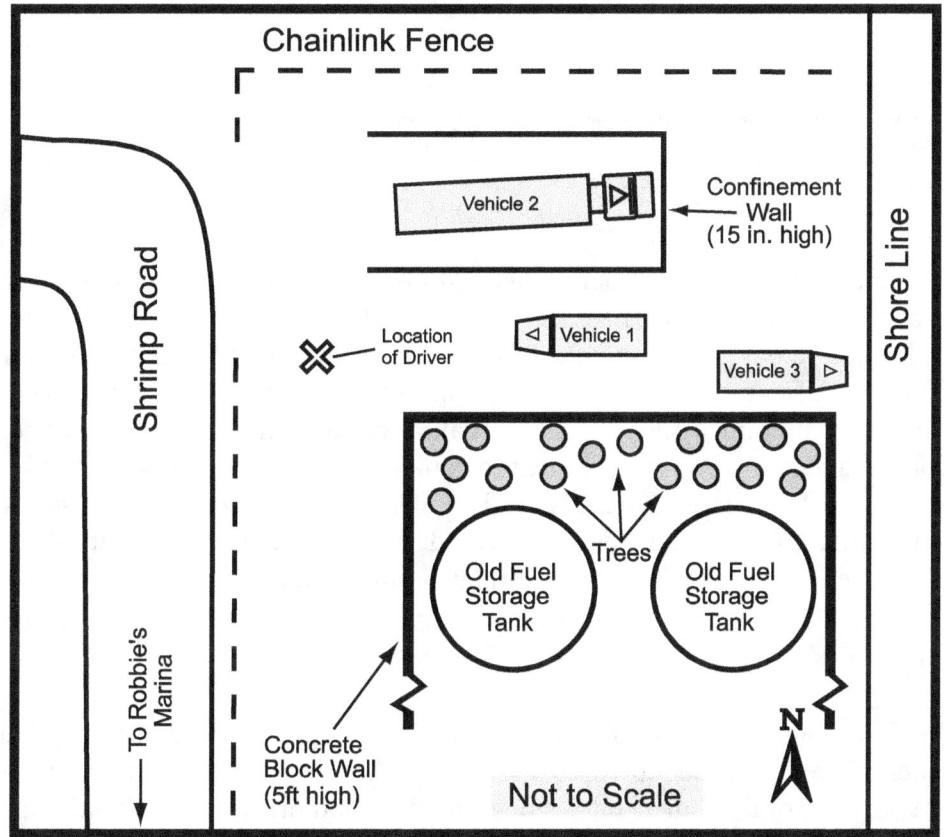

Figure 2. Accident Site (Vehicle 1 is the straight-truck cargo tank used by the driver. Vehicle 2 is the tractor semi-trailer cargo tank used as a storage tank. Vehicle 3 is the second straight truck.)

Vehicle 1, the truck used by the injured driver, had a 5,000-gallon aluminum cargo tank attached to its frame. The cargo tank was a U.S. Department of Transportation (DOT) specification DOT-406 cargo tank, manufactured in 1996 by New Progress, Inc. (now known as AmeriTank-New Progress, Inc.). The cargo tank's maximum allowable working pressure was 3.3 psig, and its test pressure was 5.0 psig. The cargo tank had three internal compartments. Its total capacity was 5,000 gallons, 3,000 gallons in the front compartment and 1,000 gallons in each of the other two compartments.

Vehicle 2, the temporary storage tank, was a combination vehicle, including a tractor and a 9,200-gallon semitrailer cargo tank. The cargo tank was a DOT specification MC-306 aluminum cargo tank manufactured in 1987 by Fruehauf Trailer Corporation. The cargo tank had five internal compartments. Table 1 shows the capacity of each compartment.

Table 1. Capacity of Compartments

Compartment	Capacity (gallons)
1 (front)*	3,100
2	1,250
3	1,300
4	1,050
5 (back)*	2,500
Total	**9,200**

* Compartments 1 and 5 had internal baffles.

Vehicle 3 was a straight truck with a 4,400-gallon aluminum cargo tank. The cargo tank had been manufactured by Eastern Tank Corp, Inc. (now known as AmeriTank-New Progress, Inc.). The date of manufacture is unknown. Dion had limited information about the cargo tank. Also, the tank specification plate was not found after the accident. The cargo tank had two compartments, a 2,000-gallon compartment in the front and a 2,400-gallon compartment in the rear. The cargo tank was neither used nor certified for the transportation of hazardous materials. It had last held a mixture of waste water and petroleum distillates that had been recovered from a customer's boat. Although the mixture had been removed before the accident, some residue remained.

Damage

Vehicle 1 sustained extensive fire damage. The upper-shell casing, the baffles, and all four heads were melted down to the chassis level in many areas. The remaining aluminum portions of the tank shell were white, dull, and brittle. Only the aluminum walls between the three compartments and a few inches of the aluminum baffle in the first compartment remained. About a third of the rear tank head remained. (See figure 3.) The remaining aluminum walls at the rear of the tank showed some evidence of tearing and bulging.

Parts of the lids and the integral relief valves (internal steel springs) for the front and middle compartments were within their respective compartments. The lid of the back compartment had been thrown about 25 feet from the vehicle. A postaccident metallurgical examination at the Safety Board's laboratory revealed that the lid had been damaged by the impact but not by heat or fire. The examination also revealed that the weld material on the flange of the lid had failed from overstress and showed no evidence of a preexisting defect.

Vehicle 2 sustained extensive damage in the front compartment of the tank and in the tractor. (See figure 4.) The head, wall, and first baffle of the front compartment were melted to about half of their original height. The second baffle was intact but had some bulging in the lower half. The remaining section of the compartment had a heavy layer of soot. The remainder of the cargo tank sustained minimal visible damage; the labels on the side of the tank facing Vehicle 1 showed some signs of soot and burning.

Figure 3. Damage to Vehicle 1

Figure 4. Damage to Vehicle 2

Vehicle 3 had melted extensively at the rear, which was facing Vehicle 1, and showed bulging and tearing in the forward part of the cargo tank. The left side of the front half of the cargo tank had been sheared from the bottom seam and pushed back onto its right side. (See figure 5.) The front baffle remained. Sections of the tank shell had multiple punctures. One of the manhole assemblies from the cargo tank was about 750 feet away, on the opposite side of the empty above-ground storage tanks, which are shown in figure 2. The assembly showed signs of having been damaged by the impact and was broken into several pieces. The left side of the truck cab and the rear supports of the cab frame had been torn, and the cab was bowed forward.

Figure 5. Damage to Vehicle 3

The concrete pads and confinement walls under and around Vehicles 1 and 2 were also damaged. Postaccident examination revealed spalling[2] on both concrete pads and on areas of the confinement wall. The spalling was more pronounced on the pad under Vehicle 1. The concrete on the confinement wall was white and chalky.

Driver Information

The driver, a 36-year-old male, had a valid Florida commercial driver's license (CDL) with a cargo-tank and hazardous-materials endorsement and an expiration date of

---

[2]According to the National Fire Protection Association, *spalling* is the chipping or pitting of concrete. Spalling of concrete can be caused by a number of things, including extended exposure to heat or fire.

November 12, 2000. He had obtained a Medical Examiner's Certificate on March 18, 1997, and it was still valid.

The driver had been driving cargo tanks and delivering gasoline and diesel fuel for about 11 years, nearly 3 years for Dion and about 8 years for Coen Oil Company in Pennsylvania.

He said that he was a smoker and that he occasionally smoked while driving his truck. He also said that he never smoked while loading or unloading his vehicle and that he was not smoking when he checked the compartments on his truck on the morning of the accident.

Carrier Information

Dion is an intrastate private motor carrier that for over 50 years has hauled bulk liquid-petroleum products, primarily gasoline and diesel fuel, in cargo tanks for local deliveries. Its main customers are commercial fishing vessels. Dion also makes deliveries to Coast Guard vessels on a call-for-delivery basis. Dion owns three tractors, four semitrailer cargo tanks, and seven straight trucks; and eight of its employees are qualified drivers.

Neither Dion nor its drivers kept records about which petroleum products were in which compartments of their trucks. The drivers controlled the loading of the cargo into their trucks and, depending on whether gasoline or one of the diesel fuels was to be delivered, were often required to switch load products in a compartment. (To *switch load* is to put a Class II or III liquid into a tank or compartment that has previously held a Class I liquid. How the classes of liquids are defined is explained table 2 below.) The driver of Vehicle 1 stated that on the morning of the accident he did not know what was in each compartment of his cargo tank.

Table 2. Classes of Flammable Liquids According to the National Fire Protection Association (NFPA)*

Class	Characteristics	Other Information
Class I	a flash point below 100° F and a vapor pressure that does not exceed 40 psia at 100° F	gasoline is in this classification
Class II	a flash point at or above 100° F and below 140° F	most diesel fuels are in this classification
Class III	a flash point at or above 140° F and below 200° F	

* Based on the NFPA's book of standards, *NFPA 385 Tank Vehicles for Flammable and Combustible Liquids* (the NFPA 385).

After the accident, Dion estimated that Vehicle 2, the temporary storage tank, had held 2,354 gallons of high sulfur diesel in the front compartment, a residue of high sulfur diesel in the second compartment, a residue of premium unleaded gasoline in the third compartment, 1,000 gallons of low sulfur diesel in the fourth compartment, and a residue of high sulfur diesel in the back compartment.

According to Dion, both compartments of the cargo tank of Vehicle 3 had held an undetermined amount of residue of waste water and petroleum distillates.

Operating Procedures—Dion had no written cargo-transfer procedures before the accident. However, after the accident the company summarized the procedures its drivers typically used as follows:

> Transfer from Dion Transport to Tank Wagon—Transport is positioned on concrete pad [cargo-transfer area] with bottom fills [bottom loading adapters] facing truck to be loaded. Truck to be loaded is positioned next to the transport in the same manner and engine is turned off. Dion driver checks transport manifest to determine amount and type of fuel in each compartment. Dion driver checks inventory on truck to be loaded to see which compartment(s) it will hold. Transport driver connects hoses from transport compartment [piping] to transfer pump. Dion driver connects output hose from transfer pump to appropriate compartment [piping] on Dion truck. All hose connections and valves are checked to make sure they are locked and in the right position. Dion driver stands between the trucks in view of the transport driver and all the connections. Transport driver starts transport and activates P.T.O. [transfer pump]. All connections are observed for leaks. While transfer is taking place, drivers are in view of all hoses and within reach of emergency shutoff. When transport compartment is empty, driver walks any remaining fuel from the hose. P.T.O is disengaged, all valves are closed, and hoses are moved to next compartment to be transferred. All compartments are off loaded in the same manner until transport is emptied. After completing transfer, all hoses are emptied and stowed and valves are rechecked. Dion driver checks site to make sure all equipment is stowed and site is clean.

Dion did not have procedures that drivers had to follow in handling the buckets of mixed fuels under the storage tank and did not require the drivers to ground and bond the cargo tanks. (For explanations of grounding and bonding, see "Static and Static Protection" in this report.) Also, the cargo-transfer area did not have a grounding rod or other means of grounding the temporary storage tank. Further, neither Vehicle 1 nor the storage tank had a bonding strap. In addition, the driver stated that he did not bond his vehicle while transferring cargo. The hose he used to transfer the product had embedded electrical conducting wires. However, the 1996 *Hose Handbook*, published by the Rubber Manufactures Association, does not recommend using such wires for bonding because "an internal static wire could break or lose contact with the couplings and not be detected visually."

Training—Dion trained its drivers in handling hazardous materials with an American Trucking Association's training program. The program includes a training video that states, "This information is for the most part basic and does not reflect the requirements particular to your company...." The program covers the following topics: classes of hazardous materials, shipping papers, labeling, loading and unloading hazardous materials (primarily small packages), placarding, and emergency preparedness and response. The program does not explain how to load and unload cargo tanks. Dion trained its drivers to load and unload cargo by having them work with an experienced operator.

Hazardous Materials

Gasoline—is a DOT hazard Class 3 (flammable liquid) with a flash point (closed cup) of -45° F, an auto ignition temperature of 500° F, and a flammable limit range of 1.4 percent to 7.6 percent in air. Gasoline has a very high volatility, corresponding to its low flash point. The NFPA considers gasoline a Class I flammable liquid.

High sulfur diesel fuel—is a DOT hazard Class 3 (flammable liquid)[3] with a flash point range of 110 to 180° F; its flammable limit range has not been determined. It has a low volatility, corresponding to its high flash point. The NFPA considers diesel fuel with this flash point range a Class II flammable liquid.

Low sulfur diesel—is a DOT hazard Class 3 (flammable liquid) with a flash point of 125 to 185° F; its flammable limit range has not been determined. It has a low volatility, corresponding to its high flash point. The NFPA considers diesel fuel with this flash point range a Class II flammable liquid.

Sources of Ignition

Static and Static Protection—Static electricity is associated with the presence of an electrical charge on the surface of an insulated body and is generated through mechanical work (i.e., scraping or rubbing of two surfaces). The generation of static electrical charges is common during the transfer of such low-conductivity fluids as gasoline and diesel fuels. Further, static charges are generated when the liquid comes in contact with another material, for example, when the liquid flows through a pipe or is mixed, poured, pumped, or agitated. Static production is likely when a flammable or combustible liquid is transferred from a container that has insulating properties and the charge is not allowed to dissipate. Plastics and rubber are common insulators used in transferring flammable and combustible liquids. If sufficient charge is produced and there is no way to dissipate the charge, a static spark can become an ignition hazard.

Static charge can be controlled so that it does not become a source of ignition. The first way is to control the flammable vapor mixture within the tank so that it does not reach the flammable range. If the mixture cannot be controlled, grounding and bonding can dissipate or "relax" the static charge. To ground a conductive object is to connect it to the ground so that any static charge is directed away from the object and into the ground. To bond a conductive object is to link it to another conductive object with a bonding wire or strap, thus reducing static sparking because both conductive bodies are brought to the same electrical potential.

---

[3]Flammable liquids (like diesel fuel) having a flash point of over 100° F can be reclassified as a combustible liquid (49 *Code of Federal Regulations* 173.120[b][2]).

According to the NFPA, switch loading either Class II or III liquids into a container that held a Class I liquid presents an unacceptable hazard. The NFPA's book of standards that applies is the *NFPA 30 Flammable and Combustible Liquids Code* (the NFPA 30). According to section A-5-6.3 of the NFPA 30:

> The term *switch loading* describes a situation that warrants special consideration.
>
> When a tank is emptied of a cargo of Class I liquid, there is left a mixture of vapor and air, which can be, and often is, within the flammable range. When such a tank is refilled with a Class I liquid, any charge that reaches the tank shell will be bled off by the required bond wire. Also, there will be no flammable mixture at the surface of the rising oil level because Class I liquid produces at its surface a mixture that is too rich to be ignitable. This is a situation commonly existing in tank vehicles in gasoline service. If, as occasionally happens, a static charge does accumulate on the surface sufficient to produce a spark, it occurs in a too-rich, nonignitable atmosphere and thus causes no harm.
>
> A very different situation arises if the liquid is "switch loaded," that is, when a Class II or Class III liquid is loaded into a tank vehicle that previously contained a Class I liquid.
>
> Class II or Class III liquids are not necessarily more potent static generators than the Class I liquid previously loaded, but the atmosphere in contact with the rising oil surface is not enriched to bring it out of the flammable range. If circumstances are such that a spark should occur either across the oil surface or from the oil surface to some other object, the spark occurs in a mixture that can be within the flammable range, and an explosion can result.
>
> It is emphasized that bonding the tank to the fill stem is not sufficient; a majority of the recorded explosions have occurred when it was believed the tank had been adequately bonded. The electrostatic potential that is responsible for the spark exists inside the tank on the surface of the liquid and cannot be removed by bonding. Measures to reduce the change of such internal static ignition can be one or more of the following:
>
> (a) Avoid spark promoters....
>
> (b) Reduce the static generation by one or more of the following:
>
> 1. Avoid splash filling and upward spraying of oil where bottom filling is used.
>
> 2. Employ reduced fill rates at the start of filling through downspouts until the end of the spout is submerged. Some consider 3 ft (0.914 m) per sec to be a suitable precaution.
>
> 3. Where filters are employed, provide relaxation time in the piping downstream from the filters. A relaxation time of 30 sec is considered by some to be a suitable precaution.

Eliminate the flammable mixture before switch loads by gas freeing or inerting.[4] [Footnote added.]

The American Petroleum Institute also recognizes the hazards of switch loading low vapor pressure products (NFPA Class II or Class III liquid) into a vessel containing a flammable vapor product (NFPA Class I liquid).[5]

Other Ignition Sources—According to Global Atmospherics, Inc., *FaultFinder Lightning Report,* lightning did not strike within 5 miles of the accident site. The closest strike was 59.5 miles away.

According to the NFPA, the temperature of the tip of a free-burning cigarette is between 930 and 1,300° F; if the cigarette is being puffed, the temperature is between 1,520 and 1,670° F.

Police video recordings of the accident revealed that the headlights of Vehicle 1 were on. The driver said that if the headlights were on, the truck's engine was running. The exhaust stack for the engine was at the back of the cab on the passenger side, more than 5 feet from the lid of the front compartment. The exhaust temperature for the type of engine that Vehicle 1 had is, when the engine is idling, approximately 300 to 400° F.[6]

Federal Regulations

The pertinent regulations are the "Hazardous Materials Regulations" (the HM regulations), 49 *Code of Federal Regulations* (CFR) Subchapter C.

According to the HM regulations (49 CFR 172.702):

> A hazmat employer shall ensure that each of its hazmat employees is trained in accordance with the requirements prescribed in this subpart ["Training,"]...and tested....

Also according to the HM regulations (49 CFR 172.704[a][2]):

> Each hazmat employee shall be provided function-specific training concerning the requirements of this subchapter or exemptions issued under subchapter A ["Hazardous Materials and Oil Transportation"] of this chapter, which are specifically applicable to the functions the employee performs.

---

[4]*Gas freeing* is evacuating all flammable vapors from the container; *inerting* is displacing all flammable vapors from the container with an inert gas.

[5]American Petroleum Institute Recommended Practice 2003, Fifth Edition, *Protection Against Ignitions Arising Out of Static, Lightning, and Stray Currents*, Section 2.4.5, December 1991.

[6]The information comes from a representative of Alban Tractor (an authorized Caterpillar dealer) in Elkridge, Maryland.

The HM regulations require that each hazmat employee receive recurrent training at least once every 3 years and that records be kept of the training each hazmat employee receives (49 CFR 172.704 [c] and [d], respectively).

Under the HM regulations, each person who operates a cargo tank is to be trained in loading and unloading procedures (49 CFR 177.816[b]), but a CDL can be used to satisfy certain training requirements (49 CFR 177.816[c]). However, a January 1997 "On Guard" bulletin prepared by the Federal Highway Administration (FHWA) states:

> (4) A CDL endorsement merely indicates that the holder has passed a minimal knowledge test concerning the area covered by the endorsement. (5) It is incumbent upon the prospective employer of a commercial vehicle driver to ensure that driver is properly trained. ...Although the CDL Tank (T) and Hazmat (H) endorsements **may** satisfy part of the hazmat training requirements of Title 49 CFR Part 172 (subpart H), possession of these endorsements does **not** relieve the employer of the responsibility for providing Hazmat training.

The HM regulations require that a cargo tank be bonded when it is being loaded through an open filling hole but not when it is being loaded through a vapor-tight top or bottom connection (49 CFR 177.837[c]). The HM regulations (49 CFR 177.834[c]) state, "Smoking on or about any motor vehicle while loading or unloading any...Class 3 (flammable liquid)...is forbidden."

State Regulations

Florida had incorporated the HM regulations in a State statute before the accident and had, substantially, the same requirements for shipping papers, specification packaging, placarding, marking, labeling, training, emergency response, and emergency telephone numbers.

Florida had not adopted the Federal regulation requiring CDLs, but had enacted its own State law, which the FHWA had accepted. Florida's *Commercial Driver License Manual for Truck and Bus Drivers* provides information on driving a tractor and cargo tank. It provides very limited information about how to load and unload hazardous materials into and from cargo tanks. Section 7-5 of the manual states, "Ground a cargo tank correctly before filling it through an open filling hole." Also, section 7-4 states, "**No Smoking.** When loading hazardous materials, keep fire away. Don't let people smoke nearby. Never smoke around...**FLAMMABLES**."

Before the accident, Florida had incorporated the NFPA's standards about fire prevention into its State law by reference.

The NFPA 385 (chapter 6) explains the requirements for operating tank vehicles that carry flammable and combustible liquids:

To prevent a hazard from a change in flash point of liquids, no cargo tank, or any compartment thereof, that has been utilized for Class I liquid shall be loaded with Class II or Class III liquid until such tank or compartment and all piping, pumps, meters, and hose connected thereto have been completely drained. A tank, compartment, piping, pump, meter, or hose that does not drain completely shall be flushed at the loading point with a quantity of Class II or Class III liquid equal to twice the capacity of piping, pump, meter, or hose, to clear any residue of Class I liquid from the system.

For flammable and combustible liquid storage tanks, the NFPA 30 (chapter 2, section 2-5.8.2.4) requires bonding or grounding for all equipment, including tanks, machinery, and piping, in which ignitable mixtures may be present. About tank vehicles, the NFPA 30 (chapter 5, section 5-6.2) states:

Bonding…shall not be required:

1. Where tank cars and tank vehicles are loaded exclusively with products that do not have static accumulation properties, such as asphalts (including cutback asphalts), most crude oils, residual oils and water-soluble liquids;

2. Where no Class I liquids are handled at the loading facility and where the tank cars and tank vehicles loaded are used exclusively for Class II or Class III liquids; and

3. Where tank cars and tank vehicles are loaded and unloaded through closed connections.[7]

According to the NFPA, since vehicles used for switch loading do not meet the above requirements, bonding is required. Appendix A of the NFPA 385 also recommends bonding to the storage tank down spout, steel loading rack, or other piping for transfer vehicles that top load liquids. Bonding just to the fill stem is not sufficient. In addition, the appendix recommends the reduction of splashing and upward spraying of liquid for vehicles that bottom load. Conductive objects, such as temperature probes not approved for use in explosive environments, should be kept away from the surface of the liquid. Flammable vapor mixtures should be freed or inerted before switch loading.

Federal/State Oversight

At the time of the accident, the FHWA had some limited authority to inspect intrastate carriers for their compliance with the HM regulations, including the financial responsibility requirements, but did not have the authority to inspect the carriers for their compliance with the requirements about training drivers and about loading and unloading cargo tanks.

---

[7]The American Petroleum Institute defines a closed connection as a connection "in which contact is made before the flow starts and is broken after the flow is completed."

In October 1998 (4 months after the accident), an amendment to the HM regulations became effective;[8] under the amendment, all intrastate shippers and carriers had to comply with the HM regulations. The Safety Board had commented on the notice of the proposed rulemaking in October 1993 and had supported the requiring of equivalent levels of safety for interstate and intrastate movements of hazardous materials by motor vehicles. The Board said it had addressed the need to improve safety requirements for intrastate motor-carrier cargo tanks in its report about a cargo tank that overturned in California in 1991, releasing gasoline and causing a fire.[9] The investigation found that California had not adopted Federal regulations beyond those in effect on October 1, 1988. As a result, in California, intrastate shipments of hazardous materials could be transported in cargo tanks that failed to meet the amended and improved Federal performance standards.

An FHWA inspection of Dion in 1994 covered only the company's financial responsibility. Because Dion was an intrastate motor carrier, the FHWA did not have the authority to inspect Dion's training of drivers in or procedures for unloading and loading cargo tanks.

At the time of the accident, Florida's motor-carrier oversight program focused on roadside inspections of vehicles and drivers. Florida inspected three of Dion's trucks in June 1998 and found each to be in violation because of one or more problems with placards, medical certificates, windshields, lights, tires, or emergency equipment. Before the accident, Florida had also audited a limited number of motor-carrier facilities for compliance with the State's hazardous-materials regulations, but it had not audited Dion. Florida has recently expanded its program and will audit more motor-carrier facilities. The program now has a senior officer and seven inspectors who will concentrate on inspecting motor-carrier facilities. The program is also attempting to use vehicle-registration data to identify all motor carriers in the State.[10]

In their investigation of the accident, Safety Board investigators reviewed the FHWA's training programs for Federal and State inspectors who inspect the facilities of highway motor carriers. The investigators found that the inspectors are trained to evaluate the function-specific training received by hazardous-materials employees but are not specifically taught how to evaluate the training programs designed to train employees in how to load and unload cargo tanks.

Fire Safety—The Florida State Fire Marshal establishes minimum fire safety standards for the State concerning storage tanks for flammable liquids, including the

---

[8]Amendment, Docket No. HM 200, *Federal Register,* Vol. 62, No. 1208, dated January 8, 1997. Docket HM-200 was developed in response to the Hazardous Materials Transportation Uniform Safety Act of 1990 (Public Law 101-615—November 16, 1990), which required the Secretary of Transportation to "issue regulations for the safe transportation of hazardous materials in intrastate, interstate, and foreign commerce" (Section 105 [a][1]).

[9]*Overturn of a Tractor-Semitrailer (Cargo Tank) with the Release of Automotive Gasoline and Fire, Carmichael, California, February 13, 1991* Hazardous Materials Accident Report NTSB/HZM-91/01. (Washington, D.C.: National Transportation Safety Board 1991)

[10]As a result of deregulation in 1980, motor carriers were no longer required to register in Florida, and old records were lost.

requirement that a storage tank be grounded to dissipate static charges. Each of Florida's 600 local jurisdictions is allowed to choose one of the four fire-safety standards approved by the State. In addition, each jurisdiction must use the NFPA 101 "Life Safety Code"[11] as a minimum standard. Each jurisdiction is responsible for establishing fire safety standards that incorporate the State's minimum standards. Each jurisdiction also has the authority to inspect storage tanks to ensure their safety. In general, the Florida State Fire Marshal will not inspect storage tanks within a local jurisdiction unless there is a complaint or other matter that cannot be resolved by the jurisdiction. The Florida State Fire Marshal also oversees the training and certification of each fire inspector, including those in the local jurisdictions.

The Monroe County fire marshal's office uses the NFPA "Life Safety Code," 1994 edition, and the NFPA 30 as industry standards in evaluating the plans for storage tanks. The assistant county fire marshal stated that vehicles used as storage tanks for flammable liquids are required to meet the county's standards for storage tanks. The county inspects a storage tank when a company applies for a building permit to construct, repair, or renovate the tank. The county fire marshal's office was not aware that Vehicle 2 was being used as a temporary storage tank for flammable liquids; consequently, the tank had never been inspected. The county assistant fire marshal stated that Dion should have applied for a building permit to use Vehicle 2 as a storage tank for flammable liquids. Although Vehicle 2 was being used as a temporary storage tank, Dion's management had not realized a building permit was necessary to use the vehicle this way and had not applied for a permit.

The assistant fire marshal indicated that the Stock Island volunteer fire chief knew about the vehicles on the site and had not told the fire marshal's office. The volunteer fire chief stated that he did not know Dion was filling the smaller trucks from the semitrailer. He thought the semitrailer was being used to transport fuel and not as a storage tank. The assistant fire marshal also indicated that his office has excellent communication with the each of the volunteer fire chiefs and will be notified if they identify anything suspicious.

Department of Agriculture—The Florida Department of Agriculture checks the accuracy of fuel transfer pumps on vehicles and storage tanks and samples the quality of fuels sold. According to State law, each oil company and service station and each truck that delivers fuel to customers must be registered with the Department of Agriculture. However, there is no requirement that private storage tanks or vehicles that do not sell or deliver fuel be registered; consequently, Vehicle 2 did not have to be registered.

Dion, its service stations, and its delivery trucks were registered with the Florida Department of Agriculture.

Department of Environmental Protection—The Florida Department of Environmental Protection is charged with protecting the State's groundwater. The department tries to prevent certain spilled materials getting into the groundwater. The department regulates above-ground storage tanks with capacities above 550 gallons and

---

[11] 1985 edition or subsequent edition.

underground storage tanks with capacities above 110 gallons. A regulated storage tank is also defined as a tank that remains at a location more than 180 days; thus a highway cargo tank could fall within the definition. Such a highway cargo tank must meet the State's environmental protection requirements for storage tanks, including the requirements about piping systems and spill containment.

Facilities within Florida must register all of their regulated tanks with the Department of Environmental Protection. The inspections of the tanks are performed by local jurisdictions under State contract. According to the Consumer Protection Administrator of the Department of Environmental Protection, inspectors do not look for tank grounding attachments, but if an inspector spots an ungrounded tank he should report it to the local fire marshal.

The Consumer Protection Administrator was aware of several calls from the State Department of Transportation about vehicles being used as storage tanks. If the Department of Environmental Protection finds that the vehicle is not registered as a storage tank, the department will have the local jurisdiction inspect it.

Dion has registered several tanks with the Department of Environmental Protection. The size and quantity of the tanks indicate that they are service stations. The cargo tank used as a storage tank at Stock Island was not registered. Dion's management indicated that the cargo tank had been moved within the past 30 days (therefore, the tank would not have to be registered).

Statistics

The Research and Special Programs Administration's recent statistics (see table 3) indicate that of the reported unintentional releases of hazardous materials from cargo tanks, most releases and almost half of the injuries and fatalities associated with releases occur during loading and unloading operations. The cost of damages resulting from releases of hazardous materials is greater for incidents that occur while the cargo tanks are en route.

Table 3. Reported Hazardous-Materials Incidents Involving Highway Cargo Tanks[a]
      1993 to 1997

	Incidents	Fatalities	Injuries	Damages
Loading and Unloading	6,381	7	290	$9,917,723
En Route	1,698	44	279	$69,408,595

[a] These are incidents reported to the Research and Special Programs Administration under 49 CFR 171.16. There may be other unreported incidents.

Analysis

Propagation of Fire

The driver indicated that he was on top of his truck (Vehicle 1), that he had opened the lid to the front compartment, and that he may have been pouring the contents of a bucket filled with a mixture of gasoline and diesel fuel into the front compartment when the ignition occurred in that compartment. However, he was seriously injured and in the hospital for almost 3 months before he could be interviewed, and his memory of the events that occurred on the day of the accident was not clear. An examination of the accident site verified much of his account of the events that occurred just before the ignition.

The damage to the three vehicles and to the concrete pads indicates that Vehicle 1 was the area where the fire started. Vehicle 1 was almost totally destroyed, while the other two vehicles had less severe damage. Also, the spalling of the concrete was most extensive on the pad under Vehicle 1, indicating that it had been exposed to high temperatures for a longer period. Further, the damage to the other two vehicles was greater on the ends that were closer to Vehicle 1. Therefore, the Safety Board concludes that the damage and fires on the other two vehicles resulted from exposure to heat from the fire on Vehicle 1.

The kind of damage that Vehicle 1 sustained indicates that the fire started in the front or middle compartment. The walls of these compartments were melted down to the chassis level, whereas about a third of the rear wall of the tank was still intact. Also, an examination of the lid of the rear compartment, which was found about 25 feet from the truck, indicated that the lid was blown off the tank before the material in the compartment was involved in the fire. The metallurgical examination revealed that the weld material on the lid flange failed from overstress; nothing suggested a preexisting defect. The fact that the lid showed no evidence of fire indicates that it broke away from the compartment before the material within started to burn. Therefore, it is likely that the fire in the front and middle compartments heated the material in the back compartment, causing a rapid overpressurization and explosion in the back compartment, which broke the lid from the tank.

After the accident, the driver was found in front of his truck, about 75 feet away from the spot on the cargo tank where, according to him, he was standing when the explosion occurred. Given the extent of his burns and his broken knee, it is unlikely that he had moved very far from where he originally fell. The explosion or pressure surge necessary to propel him such a distance from the truck would have had to come from the ignition and subsequent expansion of explosive vapors within one of the compartments. Had the vapors instead ignited outside of a compartment, they would have been unconfined and thus unable to create enough of a pressure surge to throw the driver so far. Also, because no fragments or other evidence of catastrophic failure of the front two compartments was discovered, it is likely that the pressure surge occurred through one of the compartment openings. Further, had the pressure surge occurred through one of the compartment openings, the driver would have been thrown away from the center of the explosion. Therefore, given where the driver was found, it is likely that he was, as he remembered, standing on top of the vehicle in front of the opening for the compartment when the ignition occurred.

The discovery of a manhole cover from Vehicle 3 about 750 feet away and the sheared metal on the front half of the tank shell indicate that a subsequent explosion occurred when the front compartment on this vehicle failed, which happened because the residue of the materials it contained was overpressured by the radiant heating from the fire on the nearby vehicles.

Ignition Source

Dion did not have records of the quantities and types of materials in each of the compartments of Vehicle 1. However, diesel fuel, which has a flash point between 110 and 180° F, will not readily ignite at ambient temperatures. Therefore, the fire was more likely to have been caused by the ignition of gasoline vapors, which have a -45° F flash point and will, as a result, readily ignite in almost any ambient temperature. However, according to the NFPA, gasoline will not ignite unless there is enough oxygen:

> If, as occasionally happens, a static charge does accumulate on the [liquid] surface [of a tank containing only gasoline] sufficient to produce a spark, it occurs in a too-rich, nonignitable atmosphere and thus causes no harm.

Dion's drivers switch loaded materials in the compartments of their trucks as needed to make deliveries. According to the NFPA and the American Petroleum Institute, the switch loading of gasoline and diesel fuel can create dangerous conditions within a compartment. When diesel fuel is loaded in a compartment that last contained gasoline or is contaminated with gasoline, according to the NFPA:

> ...the atmosphere in contact with the rising oil surface is not enriched to bring it [gasoline vapors] out of the flammable range. If circumstances are such that a spark should occur either across the oil surface or from the oil surface to some other object, the spark occurs in a mixture that can be within the flammable range, and explosion can result.

Static electricity is a common ignition source. A static electrical charge can be generated when gasoline and diesel fuel are transferred from a container, such as a plastic bucket, that has insulating properties. Further, if the pouring causes the liquid to splash or become agitated, a static electrical charge is generated. In fact, the NFPA indicates that splash filling is a condition to be avoided when switch loading products. Therefore, the Safety Board concludes that the ignition and fire in the cargo tank were probably caused by a static discharge in a compartment on Vehicle 1 that resulted from the driver's pouring a mixture of gasoline and diesel fuel from a plastic bucket into the compartment.

Other possible ignition sources were examined: there was no lightning in the area of the explosion; the temperature of the nearby truck exhaust stack (300 to 400° F) was not sufficient to reach the auto ignition temperature of gasoline vapors (500° F); and while smoking as a possible ignition source cannot be totally eliminated, there was no evidence that the driver was smoking, and he stated that he was not smoking when the ignition occurred.

Given the circumstances of this accident, the Safety Board believes that the FHWA should issue an "On Guard" bulletin to emphasize the danger of splash filling materials into cargo compartments and of switch loading materials having flash points at or above 100° F (NFPA Class II and III liquids) into compartments that last contained materials having flash points below 100° F (NFPA Class I liquid).

Cargo Tank Loading and Unloading

Dion's operating procedures and driver training were not adequate to ensure safe cargo transfers.

Operating Procedures—Dion had no written procedures about loading and unloading cargo tanks. According to the driver, it was his normal practice to pour the contents of a bucket filled with a mixture of diesel fuel and gasoline into an open compartment on the top of his cargo tank, and that is what he was probably doing when the fire ignited. Because Dion did not have written procedures for safely handling the cargo, the driver's unsafe practice was not prohibited by the company and, as stated earlier, probably produced a static electric charge that ignited the cargo. Therefore, Dion's lack of written procedures about the hazards of pouring a mixture of gasoline and diesel fuel from a plastic bucket into an open compartment was a causal factor in this accident.

In addition, Dion had no procedures about or equipment for grounding and bonding its vehicles to prevent the accidental ignition of flammable liquids during cargo transfer. (Because the driver had not begun transferring cargo when the accident happened, the lack of grounding and bonding procedures is not directly related to the cause of this accident.) Florida's regulations and the NFPA's standards referenced therein require that storage tanks for flammable and combustible liquids be grounded and that a vehicle be bonded when NFPA Class I and Class II flammable liquids are being switch loaded. Dion's drivers frequently switch loaded gasoline (Class I liquid) and diesel fuel (Class II liquid) in various compartments of their cargo tanks. The Safety Board concludes that Dion did not have written procedures to ensure safe cargo handling, including procedures prohibiting the pouring of flammable liquids into its open cargo-tank compartments and procedures requiring the bonding of cargo tanks when flammable liquids are being switch loaded between them. Therefore, the Safety Board believes that Dion should establish written procedures for safely loading and unloading cargo tanks.

Since the accident, Dion has discontinued its cargo-transfer operations on Stock Island. The company is loading and unloading its cargo tanks from a stationary tank in Homestead, Florida.[12]

---

[12]The procedures for loading and unloading cargo tanks at Homestead were requested; however, they were not provided.

Training—The HM regulations require employers to train and test any of their employees who handle hazardous materials; the training must include function-specific training for cargo loading and unloading. Dion did not train its drivers adequately in this area. Other than on-the-job training, the only training Dion provided consisted of an American Trucking Association's training video and test that gave only a general overview of hazardous-materials transportation. The video did not specifically address the procedures for loading or unloading hazardous materials into and from cargo tanks.

Each of Dion's drivers had a Florida CDL with a "Tank" certification. According to Federal regulations, a CDL can be a substitute for general training; however, Florida's *Commercial Driver License Manual for Truck and Bus Drivers* primarily focuses on driving a tractor and cargo tank rather than on procedures for loading or unloading hazardous materials into and from cargo tanks. Neither the Florida CDL nor the American Trucking Association's video provided complete function-specific training on loading or unloading cargo tanks with hazardous materials. In summary, the Safety Board concludes that Dion did not adequately train its drivers to ensure safe cargo handling; in particular, the company did not teach its drivers about the danger of pouring flammable liquids into open compartments of a cargo tank or about the danger of switch loading flammable liquids between cargo tanks that are not bonded. Therefore, the Safety Board believes that Dion should give drivers function-specific training on the written procedures developed in conjunction with the recommendation that Dion establish written procedures for safely loading and unloading cargo tanks. The training should explain the danger of discharging static electricity when flammable liquids are poured into open cargo-tank compartments that contain explosive vapors, the danger of transferring flammable liquids between cargo tanks that are not bonded, and the danger of explosive vapors produced by switch loading gasoline and diesel fuels.

Oversight of Intrastate Motor Carriers

The HM regulations require the training and testing of drivers on the loading and unloading of cargo tanks. An employer must train his employees in handling hazardous materials safely, particularly when cargo is being loaded or unloaded. The employer must ensure that the employee has recurrent training at least every 3 years. Florida had incorporated the HM regulations in its State regulations.

The records, however, indicate that Federal and State inspectors had never evaluated Dion's training programs and procedures for loading and unloading cargo tanks. Before the accident, the FHWA had no authority to evaluate the training programs of intrastate motor carriers. Furthermore, while Florida had the authority to inspect Dion's facilities, the State focused on roadside inspections of vehicles and not on facility inspections.

Beginning in October 1998, all of the HM regulations became applicable to intrastate motor carriers, and the FHWA gained additional authority to inspect intrastate motor-carrier operations. The FHWA is working independently and in cooperation with the States to develop programs under which both the Federal and the State governments will inspect intrastate hazardous-materials motor carriers. Also, since the accident, Florida has expanded its compliance program to increase the number of intrastate hazardous-materials motor-carrier audits.

Thus far, however, the FHWA does not train Federal and State inspectors in evaluating function-specific training for employees who load and unload hazardous-materials cargo tanks. After reviewing the Research and Special Programs Administration's statistics, Safety Board investigators found that the number of unintentional releases of hazardous materials and the number of injuries that occur during the loading and unloading of cargo tanks are significant when compared to the numbers associated with cargo tanks that are en route. The Safety Board concludes that the Federal training programs for Federal and State motor-carrier inspectors do not adequately address the need for inspectors to evaluate the training that motor carriers give their drivers on loading and unloading cargo tanks. Therefore, the Safety Board believes that the FHWA should add elements to training programs for Federal and State inspectors that include instruction on determining whether motor carriers have adequate written procedures for and driver training in loading and unloading cargo tanks.

Also, given the circumstances of this accident--the fact that before October 1998 intrastate motor carriers were not subject to all of the HM regulations, the relative number of unintentional releases of hazardous materials from cargo tanks during loading and unloading, and the fact that the Federal training programs do not adequately address the need for inspectors to evaluate the training that motor carriers give their drivers about loading and unloading cargo tanks--the Safety Board concludes that there is a need to ensure that hazardous-materials motor carriers (both intrastate and interstate) have adequate procedures for and adequate driver training in loading and unloading cargo tanks. The Safety Board believes that the FHWA should evaluate the adequacy of cargo-tank loading and unloading procedures of and driver training for hazardous-materials motor carriers and require changes as appropriate.

Oversight of Cargo Tanks Used to Store Flammable Liquids

Although the Florida Departments of Agriculture and of Environmental Protection regulate and require the registration of storage tanks used to store flammable liquids, only the Florida State Fire Marshal has the authority to regulate the fire safety of these tanks, including the grounding requirements. However, while the Florida State Fire Marshal has minimum requirements for inspecting storage tanks for fire safety and ensures the training and qualification of State and local fire safety inspectors, the storage tanks are inspected by Florida's 600 local jurisdictions.

Vehicle 2 was being used as a temporary storage tank for fuel transfer; however, Dion was not aware that using the tank as a storage tank required a county building permit and, consequently, had not applied for a permit. Also, the Stock Island volunteer fire chief (who knew about the vehicles on the site) did not know that Vehicle 2 was being used as a storage tank and did not tell the fire marshal's office about the vehicles. Therefore, the

Monroe County fire marshal's office was not aware of this tank and had never inspected it for compliance with fire safety regulations.

Because highway cargo tanks, like Vehicle 2, are not readily identifiable as storage tanks, the Safety Board concludes that local jurisdictions may not know that cargo tanks may be used to store flammable liquids and that such tanks need to be inspected to ensure that they are safe for the storage and transfer of these liquids. Although there is no list of the cargo tanks that are used as storage tanks, information could be gathered using the registration lists of the Florida Departments of Agriculture and of Environmental Protection, the list of motor carriers being developed by the Florida Department of Transportation, and information provided by inspectors for the various departments and the local fire departments.

The Safety Board believes that the Florida State Fire Marshal should make all local jurisdictions aware of the circumstances of the fire and explosions that occurred on Stock Island. The Florida State Fire Marshal should also ensure that each local jurisdiction has a program to identify and inspect cargo tanks used as storage tanks for flammable liquids to be sure the tanks meet all the fire safety standards applicable to storage tanks. The Safety Board also believes that the Florida State Fire Marshal should coordinate the help that the Florida Departments of Transportation, of Agriculture, and of Environmental Protection give local jurisdictions in identifying cargo tanks used as storage tanks for the transfer of flammable liquids to be sure that the tanks meet all fire safety standards applicable to storage tanks. The Safety Board believes that the Florida Departments of Transportation, of Agriculture, and of Environmental Protection should assist the Florida State Fire Marshal in helping local jurisdictions identify cargo tanks being used as storage tanks so that the tanks can be inspected to ensure that they meet all fire safety standards applicable to storage tanks.

Additionally, because the problem of identifying and inspecting for fire safety cargo tanks used as storage tanks may not be limited to Florida, the Safety Board believes that the NFPA, the National Association of State Fire Marshals, and the International Association of Fire Chiefs should notify their members about the circumstances of the fire and explosions that occurred on Stock Island and urge them to develop a program to identify and inspect cargo tanks used as storage tanks for the transfer of flammable liquids to be sure that the tanks meet all fire safety standards applicable to storage tanks.

Conclusions

Findings

1. The ignition and fire in the cargo tank were probably caused by a static discharge in a compartment on Vehicle 1 that resulted from the driver's pouring a mixture of gasoline and diesel fuel from a plastic bucket into the compartment.

2. The damage and fires on the other two vehicles resulted from exposure to heat from the fire on Vehicle 1.

3. Dion Oil Company did not have written procedures to ensure safe cargo handling, including procedures prohibiting the pouring of flammable liquids into its open cargo-tank compartments and procedures requiring the bonding of cargo tanks when flammable liquids are being switch loaded between them.

4. Dion Oil Company did not adequately train its drivers to ensure safe cargo handling; in particular, the company did not teach its drivers about the danger of pouring flammable liquids into open compartments of a cargo tank or about the danger of switch loading flammable liquids between cargo tanks that are not bonded.

5. The Federal training programs for Federal and State motor-carrier inspectors do not adequately address the need for inspectors to evaluate the training that motor carriers give their drivers on loading and unloading cargo tanks.

6. Given the circumstances of this accident, the fact that before October 1998, intrastate motor carriers were not subject to all of the Federal "Hazardous Materials Regulations," the relative number of unintentional releases of hazardous materials from cargo tanks during loading and unloading, and the fact that the Federal training programs do not adequately address the need for inspectors to evaluate the training that motor carriers give their drivers about loading and unloading cargo tanks, there is a need to ensure that hazardous-materials motor carriers (both intrastate and interstate) have adequate procedures for and adequate driver training in loading and unloading cargo tanks.

7. Because highway cargo tanks, like Vehicle 2, are not readily identifiable as storage tanks, local jurisdictions may not know that cargo tanks may be used to store flammable liquids and that such tanks need to be inspected to ensure that they are safe for the storage and transfer of these liquids.

Probable Cause

The National Transportation Safety Board determines that the probable cause of the accident was Dion Oil Company's lack of adequate procedures and driver training, resulting in the driver's pouring a mixture of gasoline and diesel fuel from a plastic bucket into a cargo-tank compartment that contained a mixture of explosive vapors.

Recommendations

As a result of this accident, the National Transportation Safety Board makes the following safety recommendations:

to the Federal Highway Administration:

Add elements to training programs for Federal and State inspectors that include instruction on determining whether motor carriers have adequate written procedures for and driver training in loading and unloading cargo tanks. (H-99-30)

Evaluate the adequacy of cargo-tank loading and unloading procedures of and driver training for hazardous-materials motor carriers and require changes as appropriate. (H-99-31)

Issue an "On Guard" bulletin to emphasize the danger of splash filling materials into cargo compartments and of switch loading materials having flash points at or above 100° F (National Fire Protection Association Class II and III liquids) into compartments that last contained materials having flash points below 100° F (National Fire Protection Association Class I liquid). (H-99-32)

to Dion Oil Company:

Establish written procedures for safely loading and unloading cargo tanks. (H-99-33)

Give drivers function-specific training on the written procedures developed in conjunction with Safety Recommendation H-99-33. The training should explain the danger of discharging static electricity when flammable liquids are poured into open cargo-tank compartments that contain explosive vapors, the danger of transferring flammable liquids between cargo tanks that are not bonded, and the danger of explosive vapors produced by switch loading gasoline and diesel fuels. (H-99-34)

to the Florida State Fire Marshal:

Make all local jurisdictions in Florida aware of the circumstances of the fire and explosions that occurred on Stock Island, Florida, on June 29, 1998. In addition, ensure that each local jurisdiction has a program to identify and inspect cargo tanks used as storage tanks for flammable liquids to be sure the tanks meet all the fire safety standards applicable to storage tanks. (H-99-35)

Coordinate the help that the Florida Departments of Transportation, of Agriculture, and of Environmental Protection give local jurisdictions in identifying cargo tanks used as storage tanks for the transfer of flammable liquids to be sure that the tanks meet all fire safety standards applicable to storage tanks. (H-99-36)

to the Florida Department of Transportation:

Assist the Florida State Fire Marshal in helping local jurisdictions identify cargo tanks being used as storage tanks so that the tanks can be inspected to ensure that they meet all fire safety standards applicable to storage tanks. (H-99-37)

to the Florida Department of Agriculture:

Assist the Florida State Fire Marshal in helping local jurisdictions identify cargo tanks being used as storage tanks so that the tanks can be inspected to ensure that they meet all fire safety standards applicable to storage tanks. (H-99-38)

to the Florida Department of Environmental Protection:

Assist the Florida State Fire Marshal in helping local jurisdictions identify cargo tanks being used as storage tanks so that the tanks can be inspected to ensure that they meet all fire safety standards applicable to storage tanks. (H-99-39)

to the National Fire Prevention Association:

Make your members aware of the circumstances of the fire and explosions that occurred on Stock Island, Florida, on June 29, 1998, and urge them to develop a program to identify and inspect cargo tanks used as storage tanks for the transfer of flammable liquids to be sure that the tanks meet all fire safety standards applicable to storage tanks. (H-99-40)

to the National Association of State Fire Marshals:

Make your members aware of the circumstances of the fire and explosions that occurred on Stock Island, Florida, on June 29, 1998, and urge them to develop a program to identify and inspect cargo tanks used as storage tanks for the transfer of flammable liquids to be sure that the tanks meet all fire safety standards applicable to storage tanks. (H-99-41)

to the International Association of Fire Chiefs:

> Make your members aware of the circumstances of the fire and explosions that occurred on Stock Island, Florida, on June 29, 1998, and urge them to develop a program to identify and inspect cargo tanks used as storage tanks for the transfer of flammable liquids to be sure that the tanks meet all fire safety standards applicable to storage tanks. (H-99-42)

BY THE NATIONAL TRANSPORTATION SAFETY BOARD

JAMES E. HALL
Chairman

JOHN A. HAMMERSCHMIDT
Member

ROBERT T. FRANCIS II
Vice Chairman

JOHN J. GOGLIA
Member

GEORGE W. BLACK, JR.
Member

September 10, 1999

Appendix
Investigation and Hearing

The National Transportation Safety Board was notified of the accident about 11:00 a.m., eastern daylight time, on June 29, 1998. A full go-team was dispatched from Washington, D.C., to Stock Island, Florida. Member Hammerschmidt was the Board Member on scene. Investigative groups were established for hazardous materials, survival factors, fire investigation, and metallurgy.

Parties to the investigation were the Federal Highway Administration, Dion Oil Company, Monroe County Sheriff and Fire Marshal Offices, and Sea Tow Service (an environmental response company).

The Safety Board did not conduct a public hearing during this investigation.

Abbreviations

CDL	commercial driver's license
CFR	*Code of Federal Regulations*
Dion	Dion Oil Company
DOT	U.S. Department of Transportation
FHWA	Federal Highway Administration
HM regulations	"Hazardous Materials Regulations" (49 CFR Subchapter C)
NFPA	National Fire Protection Association
NFPA 30	*NFPA 30 Flammable and Combustible Liquids Code*
NFPA 385	*NFPA 385 Tank Vehicles for Flammable and Combustible Liquids*

www.ingramcontent.com/pod-product-compliance
Lightning Source LLC
Chambersburg PA
CBHW081806170526
45167CB00008B/3353

* 9 7 8 1 4 9 6 1 4 9 4 2 8 *